The Critical Examination of X-ray Generating Equipment in Diagnostic Radiology

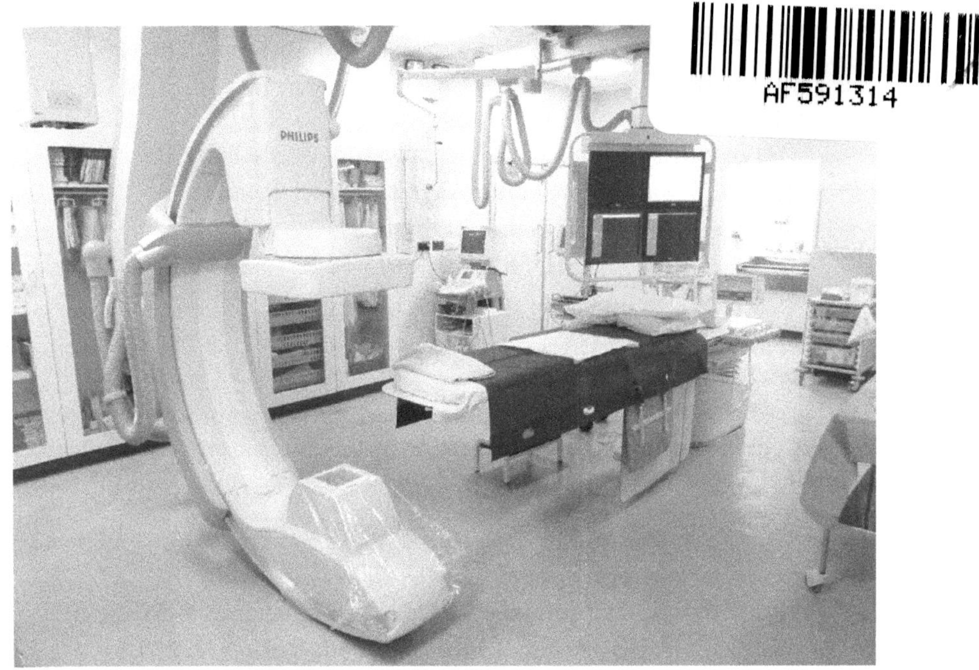

IPEM Report 107

Guidance on the interpretation of the Ionising Radiations Regulations 1999 requirements for critical examinations

Prepared by
The Institute of Physics and Engineering in Medicine
The Health and Safety Executive

Produced by a working party whose members were:
Chris Wood, Institute of Physics and Engineering in Medicine
Peter Howells, Institute of Physics and Engineering in Medicine
Gillian Rodaks, Health and Safety Executive

THIS REPORT REPLACES IPEM REPORT 79

Fairmount House, 230 Tadcaster Road
York YO24 1ES
ISBN 978 1 903613 52 8

Published by the Institute of Physics and Engineering in Medicine
Fairmount House, 230 Tadcaster Road, York YO24 1ES

Legal Notice

This report was prepared and published on behalf of the Institute of Physics and Engineering in Medicine (IPEM). Whilst every attempt is made to provide accurate and useful information, neither the IPEM, the members of IPEM or other persons contributing to the formation of the report make any warranty, express or implied, with regard to accuracy, omissions and usefulness of the information contained herein. Furthermore, the same parties do not assume any liability with respect to the use, or subsequent damages resulting from the use, of the information contained in this report.

Prepared by:
The Charlesworth Group, Huddersfield, UK. www.charlesworth-group.com

Contents

Foreword to the first report (IPEM Report 79)

The Critical Examination Working Party was set up in 1995, jointly by the Radiation Protection Topic Group (RPTG), and the Diagnostic Radiology and Magnetic Resonance Topic Group of the (then) Institute of Physical Sciences in Medicine (IPSM). The RPTG held a round table meeting in November 1994 to discuss the issues surrounding the 'critical examination' referred to in Regulation 32(2)a of the Ionising Radiations Regulations 1985. It was the feeling of the meeting that the IPSM should prepare guidance for its members on the interpretation of this regulation. This document has been published to fulfil this objective.

Membership of the Critical Examination Working Party:

Dr David Pye	Institute of Physics and Engineering in Medicine (IPEM), Nottingham (Chairman)
Mr Cliff Double	Medical Devices Agency, London
Dr Joanne Stewart	National Radiological Protection Board, Leeds
Dr David Sutton	IPEM, Dundee

Foreword to the second report (IPEM Report 545)

IPEM Report 79 was published in 1998 to provide guidance on compliance with Regulation 32(2) of the Ionising Radiations Regulations 1985 (IRR85), which introduced the requirement for a 'critical examination' by the installer of equipment used in connection with work associated with ionising radiation. This report constitutes a revision of IPEM Report 79, which is necessary due to the changes in applicable legislation. Whilst the fundamental concepts have remained unchanged between IRR85 and the Ionising Radiations Regulations 1999 (IRR99), technology has advanced considerably and updated guidance is therefore required.

Whilst IRR99 will be updated in the future, the requirement to undertake a critical examination on x-ray equipment is likely to remain. Readers are advised to refer to the particulars of current legislation, the associated Approved Code of Practice, and guidance pertaining to critical examinations, in particular the current version of HSE Guidance Note PM77 and the Medical and Dental Guidance Notes.

Mr Chris Wood, Institute of Physics and Engineering in Medicine (IPEM)
Mr Peter Howells, IPEM
Mrs Gillian Rodaks, Health and Safety Executive

The Ionising Radiations Regulations 1999, Regulation 31(2)

Where a person erects or installs an article for use at work, being work with ionising radiation, he shall –

(a) Where appropriate, undertake a critical examination of the way in which the article was erected or installed for the purpose of ensuring, in particular, that –

   i. The safety features and warning devices operate correctly; and

   ii. There is sufficient protection for persons from exposure to ionising radiation;

(b) Consult with the Radiation Protection Adviser appointed by himself or by the radiation employer with regard to the nature and extent of any critical examination and the results of that examination; and

(c) Provide the radiation employer with adequate information about proper use, testing and maintenance of the article.

Acknowledgements

The authors wish to thank all those who have provided input into this report, either through discussion or comment on drafts. In particular, thanks must go to the working party that produced IPEM Report 79 for providing such useful professional guidance and the basis for this report.

Summary points

It is the equipment installer's responsibility to ensure that a critical examination is undertaken where appropriate.

A critical examination is required following the installation or relocation of x-ray equipment, and following major service or repair of the x-ray equipment.

The need for a critical examination following service or repair depends on the nature of the work undertaken and the installer must decide if a critical examination is required.

The critical examination report should include:

- a reference to the regulation requiring a critical examination
- identification of the equipment the report applies to
- details of the tests undertaken and the results
- reference to the Radiation Protection Adviser who can be consulted regarding the results of the examination
- the name of the person undertaking the critical examination
- the date of the critical examination
- a specific conclusion about whether or not the equipment is safe to use.

Equipment must not be used clinically until the outcome of the critical examination is available.

1 Introduction

This report aims to provide clear guidance on compliance with the legislative requirement to undertake a critical examination on equipment used for diagnostic radiology. It aims to clarify the requirements of the legislation and provide advice on how to undertake and document a critical examination.

Whilst aiming to provide guidance, the authors recognise that each situation potentially requiring a critical examination is unique. As such, professional judgement will often be required, and this report aims to help individuals make that judgement.

It is hoped that this guidance will provide installers with a clear outline of their responsibilities regarding critical examinations, and as such help ensure that no item of x-ray equipment is used before radiation safety implications are considered.

Whilst this report specifically references the Ionising Radiations Regulations 1999 (IRR99), it is hoped that the principles outlined will remain relevant once IRR99 has been superseded.

1.1 Intended readership

This guidance is intended primarily for:

1. Those who have duties under the regulations to ensure that critical examinations are undertaken (e.g. equipment installers).
2. Those who undertake critical examinations on x-ray equipment on behalf of the installer.
3. Those who act as Radiation Protection Advisers (RPA) to either the radiation employer or the equipment erectors or installers.

Whilst this guidance pertains to x-ray equipment within diagnostic radiology and dental radiology, some aspects may be relevant to those installing similar x-ray equipment in other areas of work with ionising radiation (e.g. veterinary radiography). This guidance is also applicable to kV imaging devices and hybrid systems (e.g. PET/CT etc.).

1.2 Legislative background

The requirement to undertake a 'critical examination' of articles erected or installed for use when working with ionising radiation was established in the Ionising Radiations Regulations 1985 and continued in the Ionising Radiations Regulations 1999. Regulation 31(2) of IRR99 places a duty on the erector or installer of an article for use at work to undertake a critical examination to ensure that the safety features and warning devices operate correctly and that there is sufficient protection for persons from exposure to ionising radiation.

Over the years installations have become more complex in nature with an increasing number of parties being involved at various stages, including procurement and facility design. Co-operation between employers (i.e. the installer and the customer/ end user) is therefore essential to ensure that facilities are installed or constructed to specification and that they are completed in a timely manner.

To differentiate between the various radiation employers, the radiation employer purchasing or owning the x-ray equipment will be referred to as the 'customer' within this report.

1.3 Equipment life cycle

The procurement of a new diagnostic x-ray facility for medical use involves a number of stages, including the critical examination specifically required by the IRR99.

Critical examination is a distinct stage in the equipment life cycle, but may occur simultaneously with commissioning tests and/or acceptance tests. Where this overlap of tests does not take place (often in situations where the person or staff group undertaking the critical examination is not the person or staff group who would undertake the routine quality control), separate acceptance tests are still required.

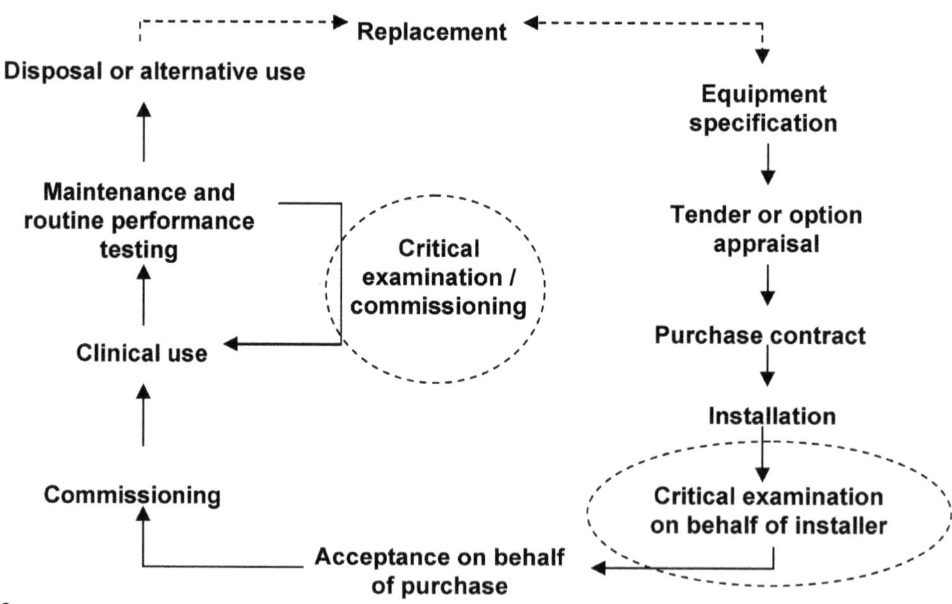

The critical examination only addresses the safety aspects of an x-ray unit and does not address equipment optimisation. Passing a critical examination does not imply that equipment meets minimum acceptability standards for medical x-ray equipment and further work may be required by the Medical Physics Expert appointed by the customer.

1.4 Radiation safety during the critical examination

The critical examination must be undertaken after installation and before clinical use. Since the critical examination addresses radiation safety issues, the person(s) undertaking the critical examination should be aware that safety features may not be in place and the equipment may not be safe to use. Professional judgement will be required to work safely using any relevant protective shielding until the safety of the equipment can be confirmed.

Particularly in the situation where a critical examination is being undertaken in a new facility, the person(s) undertaking the critical examination must also give consideration to the safety of other people in the vicinity. Cooperation between employers will be required to ensure that room shielding is sufficient before the critical examination is undertaken. The advice of an RPA may be required and professional judgement should be used to assess the particular situation.

2 Interpretation of the requirements and scope of Regulation 31(2)

This chapter aims to interpret the requirements of Regulation 31(2) in order to clarify the need for, scope of and extent of the critical examination.

2.1 The legislative context of Regulation 31

While this document considers a specific regulatory requirement placed upon the installer (i.e. the critical examination), there are other regulations that place equally important duties upon those involved in the process of bringing a piece of medical x-ray equipment into use. In particular, manufacturers of x-ray equipment also have duties under:

- The Health and Safety at Work etc. Act 1974 (which imposes general duties on manufacturers etc. as regards articles and substances for use at work).
- The Medical Devices Regulations 1994 (which makes manufacturers responsible for ensuring their products meet relevant safety requirements).

Manufacturers are reminded that duties under the above regulations must be addressed as well as those under the Ionising Radiations Regulations.

2.2 Interpretation of legislative requirements

Critical examination

An examination of the radiation safety aspects of an x-ray installation required under Regulation 31(2) of IRR99. The essence of the regulation is a critical assessment of each installation as a unique situation, addressing faults which may have occurred during equipment installation. The critical examination is intended to determine that the radiation safety of the equipment has not been compromised by the installation process.

Article for use at work

Section 53(1) of the Health and Safety at Work etc. Act 1974 defines an 'article for use at work' as being:

(a) Any plant designed for use or operation (whether exclusively or not) by persons at work, and

(b) Any article designed for use as a component in any such plant.

'Plant' is further defined such that it 'includes any machinery, equipment or appliance'.

The working party considers that software-controlled physical devices, safety features and warning devices will form part of the critical examination. In addition, the installation of modified or updated software may warrant a critical examination if there are radiation safety implications associated with the use of the software.

'...where appropriate, undertake a critical examination of the way in which the article was erected or installed...'

The critical examination addresses the way in which the article was erected or installed. As such, while mobile equipment that arrives fully assembled does not formally require a critical examination, HSE Guidance Note PM77 (3rd Edition) states that mobile equipment erected in the European Union should have been critically examined by the manufacturer. The customer should therefore be supplied with documentation by the manufacturer. The customer would still need to carry out testing of the equipment prior to clinical use to demonstrate compliance with the requirements of Regulations 8 and 32(1).

'...safety features and warning devices...'

The working party interprets this wording to refer only to the safety features which relate to exposure to ionising radiation and not those related to other potential hazards (which should be considered separately from the critical examination).

'...sufficient protection for persons from exposure...'

Paragraph 524 of the IRR99 Approved Code of Practice states that 'the requirement to undertake a critical examination requires the duty holder to consider the protection provided for people undergoing medical exposures, as well as the adequacy of protection for staff and members of the public.' As such, 'persons' is taken to mean staff, members of the public and patients.

Features of the equipment that are relevant to the control of patient dose, whether physical measures or software controlled, such as tube filtration, radiation field alignment, collimation, modes of control etc. are therefore also subject to a critical examination to ensure that sufficient protection is provided.

Other devices and features which do not form part of the x-ray equipment (e.g. walls, shields, etc.) do not need to be considered as part of the critical examination. However, examination of these features is essential to ensure that they conform to specification, and to ensure that exposures are restricted so far as is reasonably practicable, as required by Regulation 8(2).

2.3 Situations where a critical examination is required

The requirement to undertake a critical examination applies to:

- installation of equipment (whether new, second-hand or refurbished)
- relocation of existing equipment (including relocation within the same premises)
- following major service or repair work where there may be radiation safety implications, e.g. following the fitting of a replacement x-ray tube or automatic exposure control devices.

These issues are discussed further in Chapter 4.

2.4 Scope of the critical examination

The scope of the critical examination is dependent on the extent of the installation and the implications for radiation safety. A critical examination of a complete installation will require more extensive testing than the examination following replacement of a component within an existing installation. The examination following replacement of a component may not require tests above and beyond what would normally be undertaken to ensure that the replacement has been successful. In all cases, the advice of an RPA should be sought to ensure that the extent of the critical examination is sufficient.

It should be emphasised that the critical examination only concerns the radiation safety aspects of the installation and further work may be required to ensure that the equipment is operating in an optimal way. The customer should seek the advice of their appointed Medical Physics Expert. Safety aspects not related to radiation must also be addressed before clinical use.

2.5 Installer's responsibilities

The duty to ensure that a critical examination is carried out rests with the employer who erects or installs the article, not the end user.

The results of the critical examination must be available before the equipment is used clinically. Whilst a report may not be available at the time the critical examination is undertaken, some form of formal handover should take place so that there is evidence that a critical examination has been undertaken.

Whilst the critical examination is the responsibility of the erector or installer, the undertaking of the examination may be contracted to another party. Overall responsibility for conducting the critical examination still rests with the erector or installer, and any contracted work should be formally agreed and documented.

Regulation 31(2)(c) requires the installer to provide adequate information on the use, testing and maintenance of the article. It is recommended that the extent of this information is determined at the contract stage. Equipment-specific training should be supplied by the manufacturer, and the customer should ensure that adequate training is included within equipment specifications.

Whether the installer is just installing x-ray equipment or is also managing the building work within an installation (e.g. in a 'turnkey' project), the responsibility for ensuring that a critical examination is undertaken lies with the installer. This is true even if the installer sub-contracts elements of the installation. Example 1 describes a situation where this may occur.

Example 1 – Subcontracting of building works

If an x-ray equipment installer contracts a third party to undertake the installation of radiation warning lights on their behalf, the responsibility for ensuring that the warning lights function correctly rests with the equipment installer and should be included within the critical examination.

2.6 Involvement of the Radiation Protection Adviser

An RPA does not have to be present when the critical examination is carried out although they must be consulted. The RPA who is consulted may be the installer's own RPA or an RPA appointed by the customer.

The RPA's advice should be sought regarding the extent of the tests to be undertaken, and the results obtained. For simple situations, pass/fail criteria may be established which would mean that the RPA does not have to review all of the test results.

3 Contractual issues and documentation

This chapter aims to provide guidance on how to ensure the need for a critical examination is addressed at the contractual stage and offers advice on what would be expected in a critical examination report.

3.1 Contractual issues

During the tender and procurement stages of an installation, it is recommended that the end user includes and agrees at the contract stage with the erector or installer:

1. By whom the critical examination will be undertaken and the scope of the examination.
2. That the results of the critical examination will be available before clinical use, including a statement as to whether they are satisfactory, and that the equipment has been installed to specification.
3. That information is to be provided in relation to the safety features and warning devices associated to the equipment.

Advice should be sought from a RPA regarding any specific radiation safety aspects of an installation.

3.2 The critical examination report

It is expected that the following would be addressed within a critical examination report:

1. Specific reference to the requirements of Regulation 31(2) of the Ionising Radiation Regulations 1999.
2. Identification of the equipment/installation being critically examined.
3. Details of the tests carried out and the outcomes.
4. The name of the RPA who was consulted regarding the tests undertaken.
5. The name of the person carrying out the critical examination.
6. The date of the critical examination.
7. A conclusion, i.e. that the equipment has passed the critical examination.

The extent of the report will reflect the complexity of the installation and should only contain information relevant to the article installed.

The Medical and Dental Guidance Notes suggest keeping the critical examination report over the operational lifetime of the equipment.

If the outcome of the critical examination is unsatisfactory, then the failure should be reported to the customer, remedial action taken and the examination repeated. In any event the employer should not bring any equipment into use unless, if appropriate, a critical examination has been undertaken and the test results were found to be acceptable.

4 Examination of specific types of equipment

The following chapter describes some of the features that should be tested as part of a critical examination, and gives examples of situations where a critical examination may be required. Since each installation is unique, it is the responsibility of the erector or installer to determine which features are subject to examination in consultation with an RPA.

Whilst domiciliary x-ray units of the type taken into the patient's home are increasingly rare, they are specifically mentioned in HSE Guidance Note PM77 and are therefore also considered within this chapter.

4.1 General safety features

The following general safety features should be included in a critical examination (list is not exhaustive). Example 2 considers the implications of using existing safety features with new equipment.

Interlocks

- door interlocks
- emergency off buttons/switches
- beam off/disable buttons
- microswitch interlocks
- alignment and filter interlocks
- image receptor sensing interlocks.

Warning systems

- warning signals (unambiguous location, appropriate wording)
- entry warning signs (correct location, appropriate wording)
- beam on indications (minimal delay between exposure activation and indication)
- tube selection indicators (for rooms with multiple x-ray tubes)
- unambiguous labelling
- fluoroscopy duration alarms.

Safety design features

- exposure termination (by release of exposure switch)

- leakage radiation
- beam filtration and collimation
- position of exposure switch (if appropriate, can exposure only be initiated when the operator is behind the protective shield?)
- protection of exposure switch from accidental activation (especially footswitches)
- automatic exposure control (dose rate termination, back-up timer, sensor/mode selection).

Fluoroscopy-specific features

- fluoroscopy exposure termination after a fixed duration
- alignment of x-rays to fluoroscopic image receptors
- fluoroscopy dose rate limitation (e.g. the limiting values suggested in IPEM Report 91).

The appropriateness of the automatic exposure termination of fluoroscopy equipment should be considered with respect to the clinical use of the equipment. Suitable audible warnings should be in place in order to avoid the unwanted termination of exposures during an examination.

Example 2 – Use of existing safety features

If a supplier installs x-ray equipment and connects it to door interlocks and warning signs that are already in place, the supplier would have to check that the interlocks and warning signals operated correctly as part of the critical examination. Conversely, if a hospital installs interlocks and warning signals in a room that already contains installed equipment, the hospital must include the checks as part of its critical examination. Co-operation between different parties involved is essential for new installations, and also during facility modifications.

It is recommended that the advice of the equipment manufacturer is sought prior to the testing of any emergency off switches. There is a potential for the activation of such switches to cause adverse effects on equipment which may be acceptable in emergency use, but should be avoided for the purposes of testing. Confirming that powered-down equipment cannot be started with an emergency off switch activated would be considered an acceptable test.

4.2 Replacement x-ray tube inserts

X-ray tube insert replacement normally involves the disassembly and/or removal of the housing shielding, collimators, filters etc. and as such is likely to require a critical examination.

4.3 Software changes

The need for a critical examination following a change to equipment software should be determined by the radiation safety implications of the modification. Software that affects dose and/or dose rate would require a critical examination, but software dealing with patient demographics or data handling would not. The onus is on the erector or installer to determine whether or not a critical examination is necessary and to ensure that one is undertaken.

Where software has been upgraded and requires a critical examination, it is also expected that the installer should provide a report giving details of the effects of the software changes.

4.4 Mobile equipment (including specimen cabinets)

A critical examination is not required for any new mobile x-ray equipment or specimen cabinets as these are usually delivered fully assembled and have not been installed or erected. Similarly a mobile dental x-ray unit does not need a critical examination but a wall mounted x-ray unit does. Whilst these units are very similar, the distinction is explicit in HSE Guidance Note PM77 (3rd edition).

It should be noted that in situations where a critical examination is not a requirement, the functioning of safety features and warning devices should still be checked prior to first use. These checks would be the responsibility of the customer and not the installer.

The replacement of an x-ray tube insert in any mobile x-ray unit will require a critical examination, as will replacement components or changes to software if there are radiation safety implications. The onus is on the erector or installer to determine whether or not a critical examination is necessary and to ensure that one is undertaken if appropriate.

4.5 Shielding assessment

Shielding inherent to the equipment (e.g. the shielding surrounding an x-ray tube) would need to be assessed during a critical examination if there is a possibility that it may have been damaged during installation.

Shielding of a room would not normally be considered within the remit of a critical examination, and would not be the responsibility of the x-ray equipment installer.

Mammographic x-ray equipment normally features a protective lead screen as part of the operator's console. This screen could therefore be considered part of the x-ray unit and hence fall under the remit of the critical examination. As such, advice from an RPA will be required to determine what constitutes the 'sufficient protection' required by Regulation 31(2).

Additional shields (e.g. ceiling-suspended eye shields and table-mounted shields) do not form an inherent part of the x-ray equipment and therefore do not require a critical examination.

In addition, it should be noted that engineering controls such as shielding and warning devices, even if they do not require a critical examination, must be maintained and examined under the requirements of Regulation 10(1) and adequate monitoring of designated areas must be carried out to meet Regulation 19(1).

4.6 Domiciliary x-ray units

Equipment used for domiciliary radiography may require some degree of assembly before use and therefore consideration must be given to a critical examination. In this case the end user would effectively be acting as the installer and would therefore be responsible for ensuring that a critical examination is undertaken.

The scope of the tests required and what constitutes a satisfactory result should be discussed with an RPA. This will vary depending on the amount of assembly required. It is likely that a critical examination will consider:

- the beam collimation and pointing devices, and whether they are fitted correctly and are secure
- that the brakes and fasteners are secure
- that the unit is switched off before the power supply is connected to ensure that inadvertent exposures are prevented
- that all electrical connections are secure and there is no exposed wiring.

Example 3 shows a section from an example report that demonstrates a tick-list of checks that could be undertaken as part of a critical examination of a domiciliary unit.

Example 3 – domiciliary equipment critical examination tick list

Action	✓
Assemble stand making sure it is on a firm and level surface.	☐
Ensure that all brakes and fastenings are secure.	☐
Fit the tube to the stand and ensure that any safety devices are fitted and secure.	☐
Fit any collimation, filtration and pointing devices, ensuring that they are secure.	☐
Connect the unit electrically, checking for loose plugs and exposed wiring.	☐
Make sure the equipment is switched off.	☐
Connect the mains plug to the supply.	☐
Switch on and use in accordance with the equipment's instructions.	☐

Comments			
Date		Initials	

Bibliography

Health and Safety at Work etc Act 1974 (HMSO, London)

HSC (Health & Safety Commission) 2000 *Approved Code of Practice and Guidance (ACOP) Work with Ionising Radiation* (HMSO, London)

HSE (Health & Safety Executive) 1985 *The Ionising Radiations Regulations 1985.* (HMSO, London)

HSE (Health & Safety Executive) 1999 *The Ionising Radiations Regulations 1999. SI 1999 No 3232* (HMSO, London)

HSE (Health & Safety Executive) *2002 Equipment Used in Connection with Medical Exposure. HSE Guidance Note PM77 (3rd edition).* (HMSO, London)

IPEM (Institute of Physics and Engineering in Medicine) 1998 *The Critical Examination of X-ray Generating Equipment in Diagnostic Radiology. Report No. 79* (IPEM, Fairmount House, 230 Tadcaster Road, York YO24 1ES)

IPEM (Institute of Physics and Engineering in Medicine) 2002 *Medical and Dental Guidance Notes: A Good Practice Guide on all Aspects of Ionising Radiation Protection in the Clinical Environment* (IPEM, Fairmount House, 230 Tadcaster Road, York YO24 1ES)

IPEM (Institute of Physics and Engineering in Medicine) 2005 *Recommended Standards for the Routine Performance Testing of Diagnostic X-ray Imaging Systems. Report No.91* (IPEM, Fairmount House, 230 Tadcaster Road, York YO24 1ES)

www.ingramcontent.com/pod-product-compliance
Ingram Content Group UK Ltd.
Pitfield, Milton Keynes, MK11 3LW, UK
UKHW051126260726
13967UKWH00010B/2891